P9-DDF-496

THE MAJESTY OF THE RIVER ROAD

The Majesty of the
RIVER ROAD

Photography by Paul Malone
Text by Lee Malone

PELICAN PUBLISHING COMPANY
GRETNA 1988

Library of Congress Cataloging-in-Publication Data

Malone, Lee.
The majesty of the River Road / Lee and Paul Malone.
p. c.m.
Includes index.
ISBN 0-88289-674-1
1. Plantations—Louisiana—Guide-books. 2. Plantations—Louisiana—Pictorial works. 3. Louisiana—Description and travel—1981- —Guide-books. 4. Louisiana—Description and travel—1981- —Views. I. Malone, Paul. II. Title.
F370.M34 1988
917.63'0463—dc19 87-31160
CIP
(Rev.)

Manufactured in Hong Kong
Published by Pelican Publishing Company, Inc.
1101 Monroe Street, Gretna, Louisiana 70053

To the River Road plantation home owners,
who allowed us to photograph
the interiors of their homes.

Acknowledgments

Michael M. Pilié, who gave generously of his time, faith, and consideration.

Carl L. LeBoeuf, II, lab technician, who printed the photographs superbly.

Jim Calhoun and Dr. Milburn Calhoun, for their interest in the book and their suggestions.

Contents

Introduction

Along the road that follows the course of the Mississippi River stand many majestic structures that have withstood the test of time.

Vast plantations began to appear in this area of Louisiana by the 1720s. Through experimentation, it was discovered that the soil in north Louisiana was most suitable for growing cotton. In the southern part of the state sugarcane emerged as the main agricultural product. As the plantations became profitable, the owners built magnificent mansions near the banks of the river. It was important for the plantations to be near a waterway so that agricultural products could be shipped as efficiently and cheaply as possible.

The Mississippi River also made it easier for the plantation owners and their families to travel to Louisiana's social hub, New Orleans, and to other parts of the world.

The River Road was used by the planters and their families who traveled by horse-drawn carriages to each other's palatial homes, all of which reflected the prosperous and opulent lifestyle of the antebellum period.

This unique way of life ended with the Civil War. Most of the planters joined the Confederate army and left their families to fend for themselves. Many of the mansions sustained extensive damage. Some were completely destroyed by fire; others were ravaged by occupying troops.

There was a serious dearth of money when the Civil War ended. During the Reconstruction years many of the damaged plantation homes were restored. Others were left to deteriorate.

This situation prevailed until the 1940s and 1950s when many of the heirs of the original owners were able to begin restoration. Some of the homes were sold to new owners who restored them to their original splendor.

The grand and stately plantation homes, other historic structures, and the river itself renew the majesty of the River Road.

THE MAJESTY OF THE RIVER ROAD

Ashland-Belle Helene

Darrow

Ashland plantation was originally named after Henry Clay's home in Kentucky. The grand Greek Revival house, completed in 1841, was presented by Duncan Farrar Kenner to his wife, Anne Guillemine Nanine Bringier.

Kenner became one of the leading sugar planters in Louisiana. He was also a noted horseman, gambler, lawyer, and politician. In 1865 he was appointed minister plenipotentiary to Europe by Confederate President Jefferson Davis to help gain the support of France and England. An advocate of scientific methods of farming, he also founded a sugar experimental station and the Sugar Planters Association.

John Reuss purchased Ashland in 1889 and renamed the plantation Belle Helene in honor of his granddaughter, the late Helene Reuss (Mrs. W. C. Hayward, Sr.).

Built in monumental Greek Revival style, 28 three-foot-square, thirty-foot-high stuccoed brick columns support the upper gallery and a massive entablature. Surrounding the sixty-foot-square house are lower and upper twelve-foot-wide galleries.

Unoccupied and unattended from 1939 to 1946, the house suffered from the elements. In 1946, however, the Haywards began major restoration of the magnificent structure. Considerable damage was inflicted upon the home in 1959 by vandals who completely destroyed all eight Italian marble fireplaces. Ashland-Belle Helene was entered in the National Register of Historic Places on May 4, 1979.

A marble-topped bureau, a chest, and an antique bed with half tester in a guest bedroom at Ashland-Belle Helene are of carved mahogany.

A huge four-poster bed with full tester, a marble-topped bureau, and an armoire with mirrored doors furnished the master bedroom at Ashland-Belle Helene.

Note the hand-carved furniture in this bedroom in Ashland-Belle Helene.

This old slave cabin stands silently behind Ashland-Belle Helene.

René Beauregard House

Chalmette

During the 1830s this splendid Greek Revival house was built on the banks of the Mississippi River. In 1880, Judge René Beauregard, son of General Pierre Gustave Toutant Beauregard, bought the property and resided with his family in the mansion until 1904. These were happy years for the Beauregards. The general's grandchildren delighted in his visits and roamed the verdant grounds with him.

Eight huge Doric columns extend from the ground to the hipped, gabled roof in the front and the rear. An elegant wooden railing encircles both front and rear recessed galleries. Made of brick covered with cement, the house consists of two stories and an attic.

The rear of this magnificent home offers a breathtaking view of the river. The front of the home overlooks the site of the Battle of New Orleans, which took place on January 8, 1815. (The Americans, under the command of General Andrew Jackson, won a decisive victory against the British in the conflict.) Authentic cannons, reconstructed ramparts, and the Chalmette National Cemetery are visible from the front gallery.

The house is now a museum open to the public.

The front gallery of the René Beauregard House overlooks the site of the Battle of New Orleans. Authentic cannons and reconstructed ramparts are shown.

Belle Alliance

Near Donaldsonville

Built in 1846 by Charles Kock, a Belgian aristocrat, Belle Alliance served as the Kock family home until 1915. The productive sugarcane plantation consisted of 7,000 acres.

Of classical design, the front facade of the impressive house features an unusual wide stairway, which gives access to the second-story gallery. Six tremendous plaster-covered brick columns support the massive entablature, encircled by an ornamental dentil course. The balustrades on the gallery and the stairway are of intricate iron filigree.

Lush shrubbery, colorful flowers, and old live oak trees hanging with Spanish moss surround the beautiful mansion.

Bocage

Near Darrow

Bocage, which means *shady retreat* in French, majestically stands along the River Road in the midst of beautifully landscaped gardens. It was built in 1801 by Marius Pons Bringier as a wedding gift for his daughter Francoise, and Christoph Colomb, a native of Carbeille, France, and kinsman of Christopher Columbus.

After a fire in 1837, extensive remodeling took place and interior decorations in the Greek Revival style were added. The house is furnished with elegant antiques.

The two-story home of classic design employs brick construction below and cypress wood above. Across the front of the house are eight square, plastered brick pillars supporting the upper recessed gallery and massive entablature. A carved dentil course encircles the entablature. The thick outer six columns contrast with the thin and light middle pair. An *Inantis* gallery in the rear of the house is attributed to James Dakin, the architect who designed the old State Capitol in Baton Rouge, Louisiana.

In 1941, Bocage was purchased and restored by the late Dr. Edwin G. Kohlsdorf and his wife, Dr. Anita L. Crozat of New Orleans. Mr. and Mrs. Richard Genre are the present owners. Mrs. Genre, the former Marguerite Crozat of New Orleans, is the niece of the late Dr. Anita Crozat Kohlsdorf. The Genres spend as much time as possible at the plantation, welcoming their children and grandchildren from Connecticut and Virginia, as well as entertaining friends in the peace and quiet of an antebellum atmosphere.

In the front part of the central hallway at Bocage a Trumeau mirror hangs on the wall above the antique table containing a bronze statue of Satyr and Bacchante by the famous French sculptor Claude Michel Clodion, circa 1850.

In the central hallway of Bocage stands an eighteenth-century clock from France. During restoration the wooden floors were replaced with marble, and a gracefully curving stairway was installed.

Massive sliding doors separate the two parlors of Bocage. Note the Greek Revival decorations above the doors, the ceiling medallions, and the crystal chandeliers. The mantelpieces are of carved cypress. The firescreen in the front parlor is made of woolen flowers on crystal beads. The bronze Dore clock on the mantel in the rear parlor depicts Diana of the Chase. The candelabra are also bronze Dore.

Satsuma vases on the hearth in the front parlor at Bocage. The beautiful mantel set was a wedding gift to Mrs. Genre's great-grandmother.

In the dining room of Bocage is an elaborate china cabinet containing a collection of silver christening cups, lusterware, and other collector's items. A bronze Dore clock stands on the blue marble mantelpiece.

The rear view of Bocage shows the Inantis *gallery, enclosed on three sides, and a pigeonnier on either side of the house.*

Destrehan

Near New Orleans

At Destrehan one cannot escape the feeling of drifting backward in time. Built in 1787, it is the oldest plantation home in the lower Mississippi River Valley.

The house was built for Robert Antoine Robin de Logny, father-in-law of Jean Noel Destrehan de Beaupre, who purchased the plantation in 1802. During the lifetime of Jean Noel it grew to consist of 1,050 acres of land.

Indigo, corn, and rice were the first crops raised on the plantation. With an ever-increasing blight affecting indigo plants, and with the discovery of an inexpensive method for granulating large quantities of sugar, most of the planters in this area had begun planting sugarcane instead of indigo by the turn of the century.

The Destrehans and their descendants occupied the house from 1802 to 1910. They were an aristocratic family known for entertaining noted guests, among whom was the Duc d'Orleans who became the king of France. Jean Lafitte, the pirate, frequently visited the house, and his visits inspired stories that gold was hidden in the walls. It is said that on dark and stormy nights a ghostly pirate appears and points a finger at the fireplace.

Typical of the French Colonial Period, the house was three rooms wide and two rooms deep, with no connecting halls. Wide galleries surrounded the house until a garconniere was added to each side. The ground floor of brick and the upper floor of hand-hewn cypress show the influence of the West Indies. The walls are made of *bousillage,* a mixture of mud, moss, and deer hair.

Destrehan, a grand old plantation home, is now owned by the River Road Historical Society.

Massive sliding doors connect the parlor and the sitting room at Destrehan. The French beveled mirror over the white marble mantelpiece with tapestried firescreen reflects a bygone era.

The ornately carved bed and armoire in the master bedroom at Destrehan were made by Lee, a noted nineteenth century furniture maker.

During the nineteenth century, ministers, priests, and other members of the clergy traveled to the plantation homes to take care of the residents' religious needs. In the guest bedroom at Destrehan are the bishop's traveling bed and portable confessional.

This closeup view of the back of the bishop's traveling bed shows intricate carving. The bishop took his bed with him to the plantation he visited. It was made to break down into fifty-two pieces for ease in transporting and reassembling.

The bishop's beautifully carved, portable confessional at Destrehan.

The laths were removed from this section of the wall to show the bousillage, *a mixture of mud, moss, and deer hair, used in construction of the walls at Destrehan.*

The plantation bell was used to call the workers from the fields.

Wooden pegs used to hold the cypress timbers together can be seen in this attic view at Destrehan. The cypress timbers were hand-hewn on the plantation grounds.

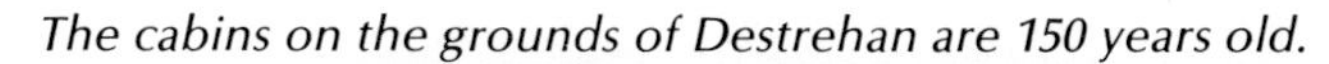

The cabins on the grounds of Destrehan are 150 years old.

Evergreen

Near Edgard

Evergreen Plantation holds several mysteries—including the identity of the architect of the house and the year of its completion. Near Edgard, the land on which this antebellum mansion stands was owned prior to 1812 by Madame Pierre Becnel II.

In 1830, Michel Pierre Becnel married Desirée Brou, and it is believed that the house was constructed about this time. Evergreen remained in the hands of the Becnel—Brou family's descendants for approximately sixty years and was a vast, productive sugarcane plantation.

Later the house fell into disrepair and continued to deteriorate until 1946 when it was purchased by Matilda Geddings Gray of Lake Charles and New Orleans. She began the painstakingly slow process of complete restoration of the manor house as well as the outbuildings.

The mansion contains ten large rooms, all of which are furnished with fine antiques. The four-poster beds feature the pineapple and acanthus-leaf motif, which is also found on the interior woodwork and throughout the grounds on gateposts.

Built in classic Greek Revival style, the house has Doric columns extending from the ground to the hipped roof, which is topped by a belvedere. The exterior of Evergreen features an unusual pedimented portico with Doric columns that match the eight columns supporting the roof. An exquisite double-curved stairway gives access to the upper gallery. Flanking the house are two pigeonniers. Garconnieres, a large kitchen, and the plantation office duplicate the fine details of the house.

Mrs. Matilda Gray Stream, the niece of Matilda Gray, presently preserves and maintains Evergreen plantation as it was before the Civil War destroyed a way of life unique to the South.

Felicity Plantation Home

Vacherie

Felicity plantation home was built by Valcour Aimé about 1846 as a wedding gift to his daughter, Felicite Emma, and her spouse, Septime Fortier.

The plantation changed hands several times between 1873 and the 1890s when it was purchased by Saturnine Waguespack. In 1901 Waguespack combined Felicity with St. Joseph plantation to form the St. Joseph Plantation and Manufacturing Company. Stan Waguespack, Sr., his son, headed the corporation.

Felicity has wide central hallways on both the upper and the lower floors, with high-ceilinged rooms opening from them. Red Italian marble mantelpieces accent several of the rooms.

Built of cypress, the two-story manor has wooden columns supporting the gabled, low-slanting roof.

Descendants of the Waguespack family still occupy this majestic mansion.

Forest Home

Near Plaquemine

The picturesque house known as Forest Home was built about 1830. John H. Randolph, originally from Virginia, had traveled down the Mississippi River and Bayou Goula from Natchez. During his journey, he noticed this quaint cottage and upon investigation found that it was owned by a Mr. Doyle. He immediately purchased the house and the vast acreage surrounding it.

Randolph planted cotton, and the plantation prospered. He and his family were content, and he affectionately called the house "my forest home." The family grew, and in 1857 Randolph began building Nottoway, a spectacular mansion on the banks of the Mississippi River. The new home was completed in 1859.

When the Civil War began, Randolph and a neighbor took their finest furniture, silver, and slaves to Texas for the duration of the conflict. Mrs. Randolph stayed at home in order to keep the plantation operating. She and her daughters faced the Federal troops when they came to Forest Home, but no damage was done to the house, although food and cattle were confiscated.

The original stairway in the well-preserved interior of the house gives access to the second story, which has a bedroom for the children's tutor and a large classroom. Separating the parlor from the dining room on the main floor are double doors containing some of the original panes of glass.

Cypress from the surrounding land was used in the construction of Forest Home. The walls were made of *bousillage* (a combination of mud, moss, and deer hair). The house is supported about three feet above ground by a hand-made brick foundation. The home is presently owned by the Forest Home Partnership.

Glendale

Lucy

An ornately carved four-poster bed in the master bedroom at Glendale. A Tiffany chandelier hangs from the ceiling.

Construction of Glendale, located on the River Road in the town of Lucy, was begun in 1805 and completed in 1807 by David Pain and his wife, Francoise Bossier. After 1840, Glendale changed ownership several times. The Lanaux family purchased it in 1922 and restored the lovely plantation home to its original beauty.

The interior of the house features carved mantelpieces detailed in the West Indies rope motif. Cypress-panelled wainscoting accents the walls. The home is furnished with antiques from the Lanaux mansion on Esplanade Avenue in New Orleans.

Although the house is two rooms deep, between the first and second rooms is an indoor window, allowing the breezes from the river to flow through the two rooms, even when the doors were closed. This was, indeed, ingenious cross-ventilation.

The recessed galleries, front and back, and the two-room depth characterize the French Colonial architecture so popular at the time the house was built.

Glendale is still a productive sugarcane plantation, and it remains in the Lanaux family.

An indoor window was built into the wall connecting the parlor with the dining room of Glendale to allow breezes to flow through. Through this window one can see an elaborately carved, burled walnut server and a plain but elegant walnut dining table.

Two marble-topped tables and graceful sofas are in the Glendale parlor. A unique piece of furniture in this room is le lit de repos, *which means "a little repose," so called because it is a daybed used for short afternoon naps.*

A most unusual armoire stands in one of the bedrooms at Glendale. Pilasters on the side of each door are on swivel hinges so that the pilasters move when the doors are opened.

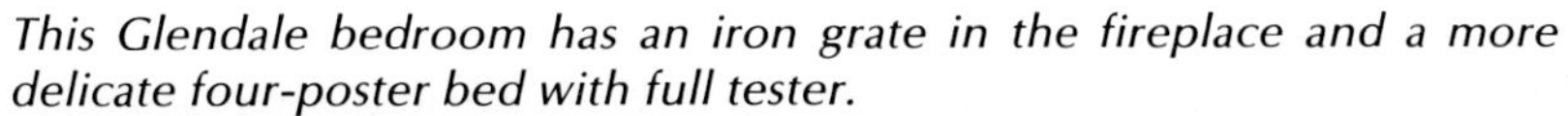

This Glendale bedroom has an iron grate in the fireplace and a more delicate four-poster bed with full tester.

L'Hermitage

Near Darrow

Near the banks of the Mississippi River stands L'Hermitage on land purchased in 1804 by Marius Pons Bringier of White Hall plantation. Bringier gave the property to his son, Michel Doradou Bringier, upon Michel's marriage to Louise Elizabeth Aglae DuBourg. Begun in 1812, the home was completed in 1814. During this period, Michel and Louise Elizabeth lived at White Hall plantation with his father.

In January, 1815, Michel served under Andrew Jackson at the Battle of New Orleans. General Jackson evidently became quite friendly with him. According to a documented account, the steamboat *Sultana* docked at the L'Hermitage wharf so Jackson could visit the Bringier family. It was after this visit that Michel named his magnificent mansion after Jackson's Nashville, Tennessee, home.

Michel died in 1847 and the vast sugarcane plantation was then managed by his wife and son, Louis Amedee Bringier.

In 1863 L'Hermitage was confiscated by the Union army. After the Civil War the family returned to rebuild the ravaged plantation and were quite successful during the dark days of Reconstruction.

Furnished with exquisite antiques, the interior of this majestic plantation home is magnificent. L'Hermitage was built in the Greek Revival style of architecture, with immense Doric columns supporting the gabled, hipped roof.

During the early 1900s ownership changed many times, and the home fell into disrepair. Dr. and Mrs. Robert C. Judice purchased L'Hermitage in 1959 and immediately began restoration. It has been restored to its original elegance and is now listed in the National Register of Historic Places.

A handsomely carved cypress mantelpiece and antique tables grace the parlor at L'Hermitage.

An antique secretary, filled with books and beautiful old china, stands in the parlor at L'Hermitage.

The sideboard and dining room table dominate this room at L'Hermitage. Both are antiques, as are the beveled mirror above the fireplace and the unique chandelier.

At the rear of L'Hermitage can be seen the old plantation bell and the exterior staircase. The staircase would have been taxed by the Spanish crown if it had been part of the home's interior.

The gazebo was an integral part of the antebellum lifestyle. The ladies would gather here to sew or read.

Homeplace-Keller House

Hahnville

With great stateliness, Homeplace overlooks the River Road. It was built for the Fortier family about 1801. Later purchased by the Kellers, it was renamed, but continued to be known as Homeplace during their occupancy.

The design of the home shows the influence of both French and Spanish architecture. As with so many other structures built close to the river, the main living quarters occupy the second floor because of the danger of periodic flooding.

Heavy round brick columns, covered with cement, support the upper gallery. Tapered wooden columns support the steep, hipped roof, which has plain but elegant dormer windows.

Towering old live oak trees shade the unoccupied mansion which seems to be sleeping—patiently awaiting a new owner.

The gracefully curved stairway at Houmas House, with delicate carving on the side of each rise, is a masterpiece of engineering.

Houmas House

Burnside

Houmas House derives its name from the Houmas Indians, who sold the land on which this house stands to Maurice Conway and Alexandre Latil. Latil built a small four-room house on the land in the last quarter of the eighteenth century.

In 1812 General Wade Hampton of South Carolina bought the plantation. The general's daughter Caroline married John Smith Preston, and they moved to Louisiana to supervise the property. In 1840 the Prestons built the magnificent Greek Revival mansion known as Houmas House. They wisely preserved the original four-room dwelling, which was later attached to the great house by an arched carriage way.

John Burnside, an Irishman, pur-

chased the plantation in 1857 and became one of the leading sugar planters in the South. He saved Houmas House from the ravages of the Civil War by claiming immunity as a British subject. In 1881 Burnside died, and the property passed to the Beirne family. It was later sold to Colonel William Porcher Miles. During his ownership the plantation flourished. After Miles died in 1899, most of the land was sold, and the grand old home fell into a state of disrepair.

The house and remaining grounds were bought in 1940 by Dr. George B. Crozat of New Orleans. He restored Houmas House to its 1840 splendor.

The furnishings in this stately manor house reflect the period prior to 1840. Many antiques came from the Crozat family home in New Orleans.

Massive Doric columns encircle the house on three sides, supporting the recessed galleries and hipped roof. The roof features graceful, arched dormer windows and a glass-enclosed belvedere.

Moss-laden live oaks, magnolias, huge azaleas, and sweet olive trees abound in the gardens surrounding Houmas House, which is superbly preserved by Dr. Crozat's heirs.

The Houmas House parlor has a warm and welcoming atmosphere. The antique furniture and the black marble mantelpiece are reminiscent of a bygone era.

A highly polished mahogany table dominates the dining room at Houmas House. Sliding doors separate the dining room and the parlor.

An antique marble-topped sideboard with hurricane lamps and a black marble mantelpiece dominate this view of the dining room at Houmas House. Above the mantel hangs an old painting in a gold-leaf frame.

This unusual view of the stairway shows its spiraling ascent from the ground floor to the third floor.

This Houmas House bedroom was used in filming the movie Hush, Hush, Sweet Charlotte. *The magnificent four-poster bed with full tester features carving in a pineapple motif.*

In the original section of Houmas House is this charming kitchen, with brass pots and pans hanging above the fireplace. A plain wooden table stands in the center of the room.

Surrounded by verdant gardens, this hexagonal brick garconniere was originally used to house overnight visitors at Houmas House. It also was used as living quarters for the young men of the family.

Indian Camp

Carville

Along the River Road near Carville stands Indian Camp plantation home, built in the late 1850s. In 1894, Dr. Isadore Dyer of Tulane University Medical School in New Orleans, needed an institution to house and treat a small group of men and women who were suffering from Hansen's disease. He succeeded in negotiating a five-year lease on Indian Camp, located eighty-five miles up the Mississippi River from New Orleans.

On the night of November 30, 1894, the patients and Dr. Dyer were transported by coal barge to the plantation. On their arrival, they found a decaying manor house and dingy slave quarters amidst magnificent live oak trees hanging with Spanish moss. The patients were housed in one of the slave cabins under the care of Dr. L. A. Wailes, the resident physician. Thus began the colony for the treatment of leprosy at Carville.

In 1896 the State of Louisiana acquired the hospital and on January 3, 1921, it was purchased by the United States government and became a national leprosarium.

Jefferson College—Manresa House of Retreats

Convent

On the banks of the Mississippi River in Convent, Louisiana, stands the main structure of Jefferson College, which was completed in 1833. The new institution enabled sons of the wealthy Louisiana planters to seek higher education without leaving their home state. Before this time the young men were sent to colleges in the North or in France. The college was named in honor of Thomas Jefferson, former President of the United States and founder of the University of Virginia.

Purchased by Valcour Aimé in 1859, the structure was closed shortly afterward because of the Civil War. The city of New Orleans fell to Federal forces in May, 1862, and five months later Federal troops occupied this splendid building as barracks.

In 1864 Valcour Aimé donated the college property to the Marist Fathers, who were citizens of France, and the United States Government ordered the troops to vacate.

Stately columns surround the main building. The magnificent, elongated structure has twenty-two impressive Doric columns in an unbroken line across its facade and a triangular pediment above the central section. It almost appears to be a replica of an ancient temple.

Presently a retreat house for laymen, this majestic, historic landmark has been renamed Manresa.

Beautiful landscaping and statuary lend an atmosphere of tranquility to the surroundings at Manresa.

This Gothic-style chapel stands on the grounds of Jefferson College—Manresa.

Kenilworth

St. Bernard

Kenilworth, in the shady serenity of its oak, pecan, and magnolia trees, is now in its twenty-second decade of existence. According to available sources, it was built in 1759 on land acquired by French and Spanish grants by Pierre Antoine Bienvenu, one of the richest and most influential men in the colony. The Bienvenu family lived in the magnificent mansion for many years.

After the original owner's occupancy, a succession of subsequent owners took place. It is said that Kenilworth once served as the temporary capitol of the French Colonial administration in Louisiana and later as the home of a Spanish Colonial governor.

Shortly after the Civil War a dinner was given here in honor of General Pierre Gustave Toutant Beauregard, and he was presented with a golden sword.

The first story of Kenilworth is built of brick faced with cement, and the second story is of cypress weatherboard. In common with most well-constructed early homes, it was built without the use of a single nail, all the parts being fastened together with mortising and wooden pegs.

Upper and lower twelve-foot recessed galleries encircle the structure. Twenty-four huge cement-covered, square brick columns support the second floor gallery, which is colonnaded with tapering pillars of slender and light wood. The galleries and exterior stairway are typical of the early

eighteenth-century plantation homes of Louisiana.

Kenilworth is rumored to house ghosts. In one legend, at the time of a full moon, a lovely wraith in a trailing white gown can be seen bending over the jasmine bushes. This is not a matter for concern, however, for Kenilworth is a place of peace, a country home where one may relax and daydream in the cool breezes. Only the sunny hours are counted.

Dr. and Mrs. Val Acosta bought Kenilworth in 1964, and they delight in preserving its elegance.

The immense columns, beamed ceiling, and batten shutters dominate the ground floor gallery at Kenilworth.

At Kenilworth, the peacock standing serenely on the exterior stairway adds to the tranquil atmosphere.

This portion of the wall in the hallway was stripped of its outer covering in order to show the briquette-entre-poteaux, *brick-between-posts, of which the eighteen-inch-thick walls at Kenilworth are made.*

The mahogany four-poster English bed, made in 1755, has an unusual reed-and-ribbon design. The lower part of the front posts are elaborately carved with bow scrolls, flowers, husks, and foliage. Beside the bed is an antique kidney-shaped mahogany table.

The top of the Vitrine table by the window opens so that small articles may be placed within. The delicately carved French dressing table with marble top was made about 1830. A Dresden chandelier hangs from the ceiling.

This view of a Kenilworth bedroom features the massive mirror, with gold-leaf frame, which hangs above the classical mantelpiece. In the corner of the room stands a stunning chair. The reed-and-ribbon motif is carried out on the chair, the table by the window, and the bed.

This closeup of the Diane de Poitiers chair shows the intricate design. Diane de Poitiers was a famous courtesan during the reign of Henri II of France.

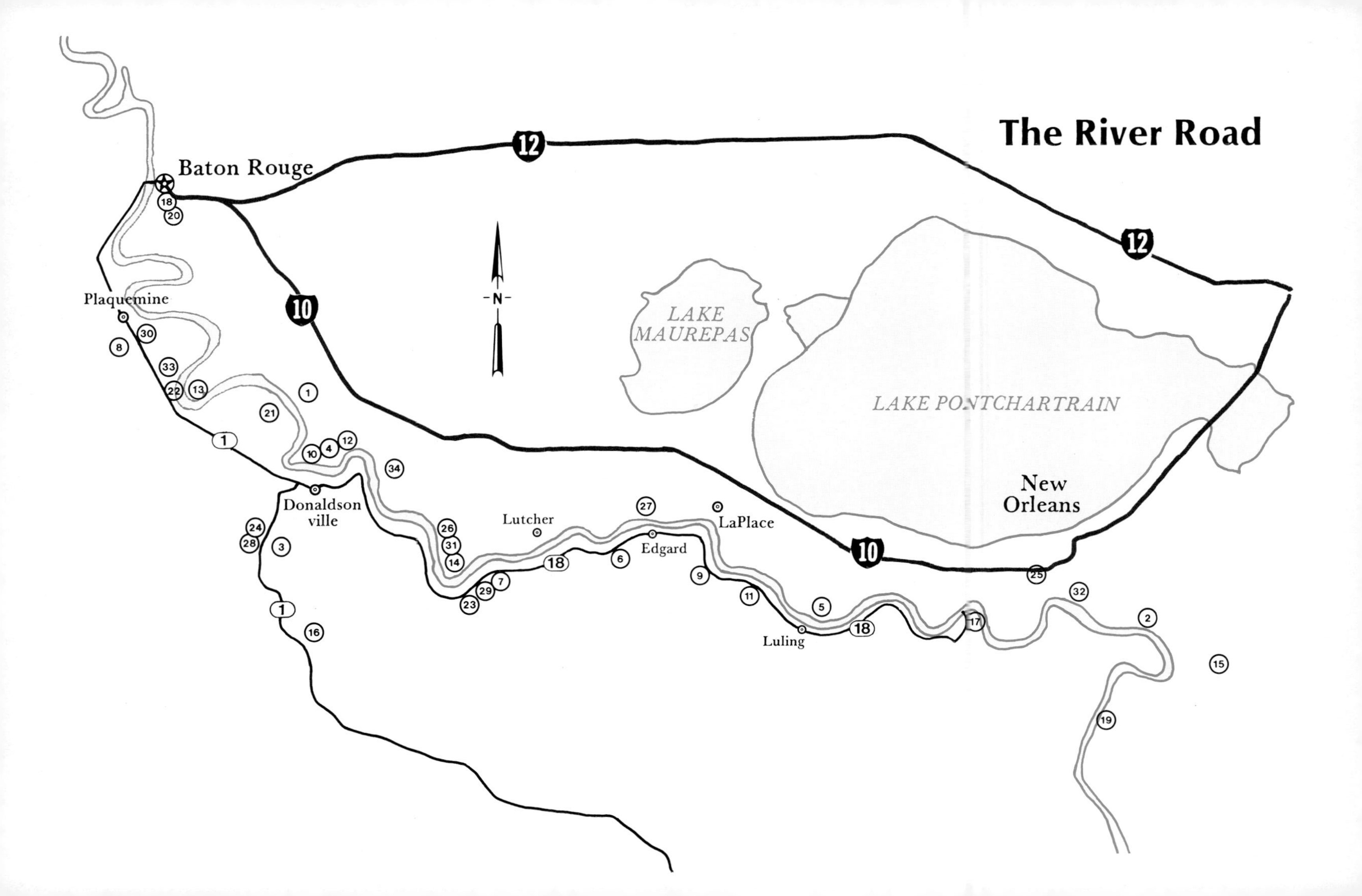
The River Road
Baton Rouge
12
10
N
LAKE MAUREPAS
LAKE PONTCHARTRAIN
New Orleans
Plaquemine
Donaldson ville
Lutcher
Edgard
LaPlace
Luling
1
18

The River Road

1 Ashland—Belle Helene
2 René Beauregard House
3 Belle Alliance
4 Bocage
5 Destrehan
6 Evergreen
7 Felicity Plantation Home
8 Forest Home
9 Glendale
10 L'Hermitage
11 Homeplace—Keller House
12 Houmas House
13 Indian Camp
14 Jefferson College—Manresa
15 Kenilworth
16 Madewood
17 Magnolia Lane
18 Magnolia Mound
19 Mary
20 Mount Hope
21 Mulberry Grove
22 Nottoway
23 Oak Alley
24 Palo Alto
25 Pitot House
26 Judge Poché House
27 San Francisco
28 St. Emma
29 St. Joseph Plantation House
30 St. Louis
31 St. Michael Church
32 The Steamboat Houses
33 Tally-Ho
34 Tezcuco

Pilasters are built into the wall alongside the graceful English walnut stairway, which has a unique landing. Note the Baccarat crystal chandelier.

Madewood
Napoleonville

In 1846 Madewood plantation house was designed by noted architect, Henry Howard, at the request of Colonel Thomas Pugh who died before completion of the magnificent mansion. Eliza Foley Pugh, his widow, directed the construction until it was finished.

A wide, central hallway on each floor, twenty-five-foot-high ceilings, and the gracefully curved stairway all contribute to the architectural excellence of Madewood. Priceless antiques can be found throughout the house, and stately Corinthian columns embellish the entrance hallway.

Cypress used in building Madewood came from the surrounding land. Bricks were hand-shaped by slaves and baked in the plantation's kilns.

The central section of the house has six Ionic columns supporting the immense entablature and pediment. A connecting wing of similar design stands on either side of the Greek Revival manor house.

Madewood plantation house was purchased in 1964 by the Harold K. Marshall family, who completely restored it to its original splendor.

Two splendid Corinthian columns and several pilasters in the foyer of Madewood.

A wedding gift to the grandparents of the present owners, the Playel piano in Madewood's music room was made in about 1900. An ornately framed, French beveled mirror hangs above it.

The center of interest in the Madewood parlor is the elegant marble mantelpiece with a firescreen to one side.

Two sideboards, one large and one small, and the antique dining room table and chairs at Madewood are made of oak in the Louisiana Renaissance Revival style.

The intricately carved bed and armoire in the bedroom are by Lee, the Cincinnati manufacturer.

At the rear of Madewood a century-old tree stands between the cemetery and an old cabin.

The cemetery at Madewood.

Under the gallery at Magnolia Lane is this carriage, seemingly waiting for a horse to be tethered to it.

Magnolia Lane
Westwego

Originally known as the Fortier Plantation, Magnolia Lane was built by Edward Fortier in 1784. The home faces the Mississippi River and the River Road (the original Old Spanish Trail). It is located on the West Bank at Nine Mile Point, so named because it is nine miles north of Canal Street in New Orleans by river.

In 1867 the Naberschnig family purchased the property and renamed it for the magnificent magnolia trees that surround it.

Influenced by the architectural style of French plantation homes in the West Indies, the main living area on the second floor is built of cypress. The ground floor is made of brick to prevent damage during the seasonal floodings of the river. About 809 panes of original glass remain in the windows and doors of the house. *Bousillage* was used in constructing the walls. Wide wooden stairs give access to the second floor galleries which surround the house. The gabled roof is double-pitched.

Magnolia Lane was the first plantation to grow strawberries in Louisiana. It also was a major nursery. Many of the largest oak and magnolia trees in the state were grown here and transplanted to other areas.

Magnolia Mound

Baton Rouge

On a high, natural ridge in Baton Rouge stands Magnolia Mound plantation house, one of the oldest wooden structures in Louisiana.

John Joyce of Fort Mobile built the home in the 1790s when he owned the surrounding cotton and indigo plantation. Joyce died in 1798, and Constance Rochon Joyce, his wife, inherited the property. Extensive alterations and additions to the home took place later when Constance married Armand Allard Duplantier. It is to this period, 1800 to 1830, that Magnolia Mound has been restored and furnished.

Bousillage was used in the construction of the walls. Strong cypress timbers, hand-hewn on the plantation, were used throughout the structure. Rooms are arranged side by side without hallways, which is characteristic of the early Louisiana Colonial style.

The home was scheduled for destruction in the late 1960s, but was rescued by concerned neighbors, the Foundation for Historical Louisiana, and the city-parish government.

A full-time director supervises a staff of local volunteers who provide narrated tours of the home, which is listed in the National Register of Historic Places.

Mary

Dalcour

This magnificent old home was built about 1773. Mary plantation passed through several owners until 1947 when it was purchased by Mr. and Mrs. E. R. Knobloch.

At the time the Knoblochs bought the home, it had been vacant for thirty-five years and had deteriorated considerably. Restoration was begun immediately, and Mary once again became the impressive West Indies-type home that it had been originally.

The interior is furnished in eighteenth-century style. Above the table in the dining room an antique wooden punkah was pulled back and forth to cool the diners.

Twenty-four Doric columns surround the house on the ground floor. On the second floor gallery, tapered wooden columns support the high pitched roof. One chimney in the center of the roof services four fireplaces within the house.

Bricks used in the lower part of the house were hand-shaped and baked in the sun. Cypress timbers were hewn on the plantation for use in the upper part of the house.

Mary stands majestically in the center of landscaped gardens and century-old oak trees.

The huge brick fireplace dominates the dining room at Mary. Note the old brick walls and the cypress punkah over the polished oak table.

The rear view of Mary plantation house is complemented by beautiful gardens.

Mount Hope

Baton Rouge

In 1786 a four-hundred-acre Spanish land grant was bestowed on Joseph Sharp, a planter who had come to Louisiana from Germany. He built Mount Hope plantation house on this fertile land in 1817.

The plantation was prosperous until the Civil War, when it served as an encampment for Confederate troops. It changed ownership several times throughout the years and deteriorated to a great extent.

In the 1970s Arlin K. Dease, one of the South's leading restoration specialists, restored Mount Hope to its original elegance. The interior has white wooden mantelpieces, mirrors, and chandeliers. The antique furniture is from the 1840s. This charming cottage, constructed of cypress dressed on the plantation by slave labor, has a recessed gallery across the front and on each side of the house. Slender columns support the hipped roof.

The present owners are Mr. and Mrs. Jack Dease.

This side view of Mount Hope shows a beautifully landscaped garden with a graceful fountain surrounded by colorful flowers.

Flowers abound at Mount Hope.

Mulberry Grove

Near Donaldsonville

Almost hidden by oak trees and shrubbery, Mulberry Grove is located on the River Road between Donaldsonville and White Castle. Built in 1836 by a Dr. Duffel, it stands back from the levee, safe from the waters of the Mississippi River.

The history of Mulberry was lost for many years because the house and land, along with other properties, were bought by John B. Reuss and grouped together under the name of Germania Plantation. Reuss, a native of Germany, acquired holdings on both sides of the river.

A daughter of Reuss inherited the Mulberry portion of Germania from her father and in 1951 sold it to Mrs. C. C. Clifton. Mrs. Clifton learned of the original name and documented the history of the 704-acre tract of land. An early map of the region, Norman's chart of 1858, shows Mulberry Grove as one of the well-known plantations along the Mississippi River.

Suffering from the abuses of many tenants and long in a state of disuse and disrepair, the house was a repository for hay when restoration was begun. It took a full year to restore this splendid manor house to its original stateliness.

Mulberry Grove reflects the French influence in this part of the country, with its first story of brick and its second story of cypress frame construction. Characteristic, too, are the wide galleries spanning the width of the

house, both front and back. Brick columns support the second-story galleries and smaller, graceful colonnettes support the third floor and the steep, sloping roof.

The magnificent home remains in the care of the Clifton family.

The walls of the wide entrance hallway of Mulberry Grove are covered with colonial-type paper. Over an antique chair hangs a century-old picture of a Tyrolean scene by Boehm. A unique chandelier and an antique lamp add to the airiness and spaciousness of the hall.

A Ming bowl on the old piano, the Chippendale sofa, and the French chairs create a warm, welcoming atmosphere at Mulberry Grove.

The elegant dining room at Mulberry Grove.

This charming Mulberry Grove bedroom has two canopied beds, an antique cradle, an oriental rug, and a Chippendale chest.

The massive four-poster bed dominates this Mulberry Grove room. Note the intricately-carved, marble-topped bureau.

Nottoway

White Castle

John Hampden Randolph commissioned noted architect Henry Howard to "build for me the finest house on the River Road." Construction was completed in 1859, and Randolph named the elaborate manor after his original home in Nottoway County, Virginia.

During the Civil War Nottoway was saved from destruction through the kindness of a Union gunboat officer who called for a cease-fire when he recognized the mansion at which he had once been a guest.

The authentic antique furnishings in the sixty-four rooms reflect the historical essence of the home. Descriptive documents were found and followed in detail during its restoration. The spectacular white ballroom, adorned with hand-carved Corinthian columns, archways, intricate plaster friezes and medallions, original crystal chandeliers, and two hand-carved marble mantelpieces is a masterpiece of beauty and function. Though cypress from the plantation land was used throughout the building, the Randolphs decided the ballroom floor should be of maple, a harder wood.

Nottoway's architectural style is a blend of Greek Revival and Italianate, with a massive entablature supported by twenty-two enormous columns. Balustrades of ornate iron filigree surround the galleries.

Visitors can feel the majesty of the South as they stand on the gallery and view the Mississippi River. Oak and magnolia trees accent the grounds around the mansion.

Nottoway was completely restored by Arlin K. Dease, one of the South's leading restoration specialists. Though this splendid manor has changed hands, the new owner constantly maintains and enhances it.

Sheer elegance is reflected in the beauty of the white ballroom at Nottoway.

This is the bright and cheerful room in which the ladies entertained while the gentlemen had coffee and cigars in the parlor after dinner at Nottoway.

The parlor at Nottoway resounded with lively conversation when the gentleman gathered there after dinner.

The music room at Nottoway.

The bedrooms at Nottoway are furnished as they were originally. Information taken from old documents found when restoration was begun specified the original colors and materials.

Tulip garden at Nottoway.

From the cool, shadowy, third-floor gallery at Nottoway one can see the Mississippi Queen *docked at the plantation's wharf. The view recalls the time when many paddlewheel steamboats plied the Mississippi River.*

Oak Alley

Near Vacherie

Oak Alley was built by Jacques Telesphore Roman, III, a wealthy sugar planter, for his bride. Construction took place between 1837 and 1839, approximately one hundred years after an unknown French settler had planted the widely known, symmetrically placed alley of oaks. The twenty-eight beautiful oak trees extend from the front of the house to the River Road. Nowhere is there a more spectacular setting for a grand mansion.

Of Greek Revival architectural style, Oak Alley has twenty-eight immense Doric columns, each eight feet in circumference. The blush-colored columns of brick covered with stucco encircle the house and support the great entablature.

Oak Alley escaped damage during the Civil War. There was a succession of owners after 1866, however, and the home was finally left to the elements. In 1925, in an advanced stage of deterioration, Oak Alley was purchased by Andrew and Josephine Stewart and became the first of the great River Road plantation homes to be fully restored.

Under the provisions of Josephine Stewart's will, the mansion became a non-profit foundation so that visitors may rejoice in its heritage.

Oak Alley is managed by Mr. Zeb Mayhew.

In the restful parlor at Oak Alley, the predominant piece of furniture is an antique Heppelwhite table made in Louisiana.

The beautiful parlor at Oak Alley is enhanced by the magnificent grand piano manufactured in 1857 by John Broadwood and Sons of London. This firm has held the Royal warrant since 1740 and in 1818 gave a similar grand piano to Beethoven.

The antique dining room table and chairs at Oak Alley are of mahogany. The mantelpiece is of hand-carved cypress with a marble border around the fireplace.

This antique four-poster bed with full tester in the master bedroom at Oak Alley was made of walnut by Mallard.

Canopied twin beds and baby's crib in the children's bedroom at Oak Alley.

Palo Alto

Near Donaldsonville

Palo Alto plantation house was built about 1850 on the banks of Bayou Lafourche, part of the broad River Road area. There is no record of the original owner; it was acquired in 1852, however, by Pierre Oscar Ayraud and his wife, Rosalie Rodriguez, from the succession of her father, Mathias Rodriguez. Jacob Lemann purchased the home in 1860.

The choice of the Spanish name Palo Alto, meaning *high tree,* may have been influenced by a battle in the Mexican War, on May 8, 1846, in which militia from Louisiana and Texas fought under General Zachary Taylor.

Palo Alto and the adjacent plantation, St. Emma, were the site of a series of Civil War engagements in which the Union lost 465 men.

The interior of the plantation home fascinates the visitor. Furnished with authentic antiques, the structure features intricate cornice and ceiling plaster work. The central hall has a cornice ornamented with Greek palmettes, below which is a cove filled with a rinceau motif in high relief. Further below, spaced rosettes punctuate the frieze. There are two leafed medallions in the entrance hall.

The main house, a one-and-a-half-story structure, stands about four feet above the ground on brick piers. Two dormer windows break the high-pitched roof on each slope. A twelve-foot-wide front gallery, recessed beneath the gabled roof, spans the width of the house. Its deep entablature returns along the depth of the gallery. Cypress from the plantation land was

used in its construction. The walls are of *briquette-entre-poteaux* (brick-between-posts). To the side of the house, which formed the original kitchen area, a similar roof and recessed gallery are duplicated.

This beautiful, comfortable old home stands surrounded by informally landscaped grounds with many large live oak, pecan, magnolia, cypress, and other trees. Still a productive sugarcane plantation and occupied by the Lemann family throughout the years, Palo Alto plantation house continues to be well-preserved.

The single-paneled entrance features unusual, octagonally-shaped, clear glazed lights in the transom and on either side. The intricate cornice work and hand-hooked rugs add to the charm of the hallway at Palo Alto.

Above the Duncan Phyfe sofa hangs a copy of a Persac painting. The piano, one of two in the parlor, was made by Beauvais.

Pitot House

New Orleans

James Pitot bought this house with its thirty-acre sugarcane plantation in 1810. He had already attained prominence in New Orleans by establishing himself as a commission merchant and serving as commissioner with the Cabildo. He later became the first democratically elected mayor of New Orleans.

The people loved and revered Pitot because of his dedication to their welfare. When he died in 1831 the courts and city council adjourned, and the people mourned the city's most distinguished and respected citizen.

The interior of Pitot House is beautifully furnished with authentic antiques of the period. Great effort was made to discover and restore the original interior design and decoration. Scrapings revealed the original paint colors, which are now used throughout the house.

Typical of the cool and comfortable West Indies style of architecture, galleries surround the house on three sides. Massive Doric columns on the ground floor and slendor colonettes above support the double-pitched roof. A decorative iron grille encloses the rear loggia, and jalousies screen the rear gallery.

When Pitot House was threatened with demolition in 1962 because the land was needed for other uses, the Louisiana Landmark Society rescued this historic home by moving it to a nearby site.

The present resident curators, Mr. and Mrs. Don Didier, have maintained the home's restoration.

This view of James Pitot's office and receiving room shows a candlestand and two eighteeneth-century saddle-seat Windsor chairs. Above the desk, made in Louisiana in 1830, hangs a portrait of James Pitot painted in 1802. The large brass candlestick was also used as a bell by holding it upside-down and hitting it with a knife. On the wall near the Hepplewhite table is a map of Louisiana, drawn in 1732 and published in 1752. A chair of Louisiana pecan with its original corn-shuck seat stands between the table and the desk. Above the mantel hangs an Eglomise*, a reverse painting on glass, of George Washington.*

James Pitot's office and receiving room. A pewter teapot and tea caddy can be seen on the eighteenth-century walnut Pembrook table. Note the cypress beamed ceiling and flagstone floor.

Four-poster twin beds of graceful design dominate the children's bedroom. An antique doll, sitting on a small chair, adds to the room's charm.

A cypress Punkah above the dining room table was pulled back and forth by a servant to keep diners cool and comfortable. The mahogany table and graceful Seignouret-type chairs date from the late eighteeneth century. An "old Paris" tureen in the center of the table and an original Audubon painting above the marble-top server augment the room's authenticity.

Judge Poché House

Convent

Built about 1870 near the banks of the Mississippi River, this plantation home served as the residence of Judge Felix Pierre Poché until 1880 when he moved to New Orleans. It later served as his summer home until it was sold in 1892.

Poché was a Civil War diarist, Democratic party leader, prominent jurist, and one of the founders of the American Bar Association. His Civil War diary is regarded as an important source of information for scholars, especially those studying the war east of the Mississippi River in the waning months of the conflict.

Priceless antiques fill the interior of the home. Each floor has a wide central hall graced by an elegant, curved stairway in the rear. The two front rooms have marble mantelpieces with large round arches and scroll keystones. Floorboards, louvered shutters, and cast-iron vents are original to the house.

The architecture of the Judge Poché house shows the influence of the Victorian Renaissance Revival Period. Built of cypress, the house has nine-foot galleries on three sides and a kitchen wing. The upper half-story is set under a broad gabled roof.

The present owners delight in preserving this extraordinary old home.

A curving stairway and an antique chest of drawers at the rear of the Judge Poché House hall.

A sparkling Baccarat chandelier lights the Judge Poché House music room. French antique furniture and an old grand piano invite one to enter and enjoy the peaceful atmosphere.

The Sheraton table in the dining room of the Judge Poché House was made about 1760. The Japanese motif of the hand-painted wallpaper accentuates the oriental motif used elsewhere in the house.

Furnished in the traditional River Road plantation house manner, the massive four-poster bed with full tester is by Prudent Mallard, the noted New Orleans furniture maker.

San Francisco

Garyville

When Edmond B. Marmillion began building his mansion called San Francisco, he planned to make it a riverside landmark, a structure to excite the admiration of passengers on the Mississippi River's steamboats.

The house was completed in 1856, the year of his death. His son, Antoine Valsin, inherited the vast sugarcane plantation and managed it profitably until his death in 1871. His widow sold the property to Achille D. Bougere in 1879.

In the luxurious style of the 1850s, the furniture was made of richly figured rosewood by noted cabinetmaker John Henry Belter. Elaborate, frescoed ceilings and painted friezes in both parlors are exquisite.

Because of the danger of seasonal flooding of the river, bricks were used for construction of the ground floor. The upper part of the house is built of cypress with wooden pegs. Heavy square brick columns support the upper floor, and graceful Corinthian columns support the immense attic and entablature. For the purpose of ventilation, louvres encircle the base of the attic.

The San Francisco Plantation Foundation now owns this fully restored structure.

In the foyer at San Francisco an unusual seating arrangement seems to invite the visitor to rest awhile and enjoy the elegant surroundings.

The downriver parlor at San Francisco. The stereopticon in the center was used for viewing pictures.

The upriver parlor at San Francisco. The Belter rosewood furniture, antique mirrors, ceiling fresco, and frieze are spectacular.

This view of the upriver parlor at San Francisco emphasizes the chandelier and the blue mantelpiece, which appears to be marble, but is crafted of painstakingly painted and marbled wood.

Elaborately draped with mosquito netting under satin, this rosewood daybed at San Francisco was used for short afternoon naps by the planter's wife.

The master bedroom at San Francisco is richly appointed in rosewood, English chintz, and velvet Brussels carpeting. The pot de chambre *surprisingly contains a water tank and flushing mechanism in the same box, or case.*

St. Emma

Near Donaldsonville

St. Emma plantation house towers over the surrounding landscape with great dignity about four miles south of the town of Donaldsonville. Charles A. Kock, a native of Bremen, Germany who owned the plantation from 1854 to 1869, became one of the leading sugar planters in Louisiana.

A nearby plantation, Palo Alto, and St. Emma were involved in Civil War skirmishes in the latter part of 1862 and early 1863. Union forces marching from Donaldsonville were checked by the Confederates in the vicinity of St. Emma, losing 465 men. The sugar houses of both plantations were used to quarter Confederate troops.

A wide central hall divides the house on each floor. There are several mantelpieces in the home, the more formal of black marble. Much of the furniture is of the Empire Period. The valances are of wood covered with gold leaf and original to the house. Though the downstairs rooms are plainer, a fine wainscotting runs the length of the hall.

Representative of a large, two-story, mid-nineteenth-century plantation house, St. Emma's lower floor is constructed of brick and the upper floor of cypress. Although the upper story contains the main living quarters, there are rooms on the ground floor which appear to be original to the house. Both front and rear facades have recessed galleries with brick and plaster pillars supporting the upper floor gallery and wooden columns supporting the hipped roof. Exterior stairways give access to the main floor.

Mr. and Mrs. Shelby Gilley, the present owners, are preserving and maintaining this stately home's restoration.

An eloquent reminder of the past, the huge gold-leaf-framed mirror in the hallway reflects a five-foot by eight-foot painting of a Major Bird of the Confederate army. The portrait was painted in Vicksburg while the city was under seige.

In the tranquil atmosphere of the parlor at St. Emma the pier mirror commands attention. A bronze Dore clock is in the center of the gold-leaf table. The gold-leaf valances on the windows are original to the house.

The impressive bed in the master bedroom of St. Emma is a unique combination of cabinetwork by Lee and Mallard. Mallard's initials were found on the back of the headboard behind the shell motif, and Lee's were on the back of the tester. A pilaster was built on the side of each door of the antique armoire.

At the rear of St. Emma is the second exterior staircase which would have been taxed by the Spanish Crown had it been built in the interior of the house. An antique carriage is beside the stairway.

St. Joseph Plantation House

Vacherie

A rear view of St. Joseph. Two exterior stairways and a simple but elegant wooden railing across the second-floor gallery indicate that it was built in the early nineteenth century.

St. Joseph plantation house is believed to have been built about 1820 by Cazamie Bernard Mericq, a native of France.

In 1858 Dr. Mericq's widow sold the house and land to Alexis Ferry, who enlarged the house and enclosed the ground floor. After the Civil War the plantation was purchased by Edward Gay, who immediately sold it to Joseph Waguespack.

The interior of St. Joseph has high ceilings and a large central hallway with rooms on each side.

Eighteen-inch-thick, solid-brick walls and brick columns supporting the second floor indicate the influence of the West Indies style of architecture. The upper part of the cypress house has ten slender columns spanning the second floor gallery and supporting a hipped roof broken by dormer windows and four chimneys.

This stately plantation house still graces the landscape facing the River Road, preserved and beloved by the descendants of Joseph Waguespack.

St. Louis

Near Plaquemine

Beautiful gardens surround St. Louis plantation house, built on the banks of the Mississippi River in 1858 by Edward James Gay, who came to Louisiana from St. Louis.

After moving to Louisiana, Gay became one of the most influential sugarcane planters in the state. He was the first president of the Louisiana Sugar Exchange of New Orleans.

Because of ill health, he did not fight in the Civil War. His son served in the Confederate army, while the elder Gay remained at home and watched the ruin and destruction that occurred. When the war ended, he directed his energies toward rebuilding the economy. After his death in 1889 the house was occupied by his widow, and later by his son Andrew.

The interior of the house has the usual wide central hall flanked on either side by large rooms, twenty-feet square. All of the main rooms on both floors have heavy cove moldings, plaster ceiling medallions, and carved Italian marble mantelpieces.

The unusual exterior of this majestic home features six Ionic columns supporting the upper gallery and six fluted Corinthian columns supporting the elegant carved entablature. The balustrades on both galleries are of intricate iron filigree.

This impressive mansion has been in the possession of the Gay family since it was built.

A large ornate pier mirror, about six feet wide and ten feet high, dominates the central hallway at St. Louis. Its frame and pilasters are of gold leaf.

In the twenty-by-forty-foot hall of St. Louis one sees a delicately carved stairway with an antique piano, made of rosewood by Mathushek in 1884. An antique clock is at the rear of the hall.

The superbly carved Italian marble mantelpiece, with a coal skuttle at its side, and the antique sofa and chairs create a warm and welcoming atmosphere at St. Louis.

Presently used for storage, the original kitchen in an asymmetrical rear wing of St. Louis contains an antique bathrub in front of the open brick hearth. Into the chimney of the hearth is plugged a wood-burning stove, circa 1865, and a very early hot-water heater. Note the old chopping block.

This beautiful china cabinet in the dining room at St. Louis contains many porcelain, silver, and crystal antique items.

This charming view of a bedroom at St. Louis has mahogany twin beds and an iron crib.

St. Michael Church
Convent

In 1809, the year of its completion, the original St. Michael Church, located approximately midway between New Orleans and Baton Rouge on the River Road, was solemnly blessed by Monsignor Jean Olivier.

Because of a great increase in population on the left bank of the river, a church in this vicinity was deemed necessary. There was no pastor at the church, however, until 1812. The parishioners' spiritual needs were ministered to by the priest from St. James Church across the Mississippi River.

In 1812 Reverend Gabriel Chambon de la Tour became St. Michael's first pastor. A succession of pastors came until 1823 when Reverend Charles de la Croix was appointed to the post. He brought dynamic zeal and sympathetic understanding to his devoted parishioners, and under his direction the struggling young parish rapidly grew and developed.

The population of the parish steadily increased, and the present St. Michael Church was built in 1831. Its architectural style is a combination of Roman and Gothic. The huge steeple, sacristies, and wings were added by 1875.

The timeless beauty of this magnificent structure is breathtaking.

The exquisite altar of St. Michael Church was constructed in 1870.

The impressive wood-beam ceiling, ornate columns, and muted colors create a feeling of serenity at St. Michael Church.

Visitors may be surprised to see this unique replica of the Grotto of Lourdes at Massabielle, France. Constructed about 1876, it is directly behind the main altar of St. Michael Church.

The Steamboat Houses

New Orleans

Two identical mansions called the Steamboat Houses stand at the end of Egania Street in New Orleans. The first was built on the banks of the Mississippi River by Milton P. Doullut in 1875. Doullut, a riverboat pilot who came to New Orleans from France, designed his home to resemble the steamboats that plied the Mississippi River at that time. The floor plan called for four spacious rooms on each floor, divided by a wide central hallway. Recessed galleries surround the house. Opening onto the galleries, floor-length windows temper the sultry climate by providing cross-ventilation.

Doullut's only child, Paul, must have been happy growing up in this ornate mansion because in 1912 he built an identical house one block away, recreating every detail of his father's house.

On the ground floor, slender Ionic columns support the second-story galleries. Delicate, yet elegant, colonnettes support the roof, which is trimmed with elaborate ironwork. Strands of large wooden "pearls" are draped around the upper galleries, and tall, tin chimneys flank the glass-enclosed octagonal belvedere.

Descendants of the Doullut family presently occupy both mansions.

At the foot of the narrow indoor stairway, original to Tally-Ho, is an antique marble-top server with original English color prints above it.

Tally-Ho

Near Plaquemine

The original owner of stately Tally-Ho Plantation House, built in the early 1800s, is unknown. Iberville Parish records show that it was bought by Jean Fleming prior to 1835. John D. Murrell of Lynchburg, Virginia, purchased the property in 1848, and it has remained in the Murrell family since that time.

Throughout the years, priceless antiques have been collected by the family. Originally only the upper floors served as a living area. Except for the kitchen, the ground level was used as a carriagehouse and storage area. The narrow stairway, giving access to the upper floor, has been enclosed.

In typical Louisiana colonial style, the lower floor is of brick, and the rest of the house is cypress. Across

the facade are upper and lower recessed galleries. Massive brick columns support the hipped roof.

Tally-Ho remains a vast, productive plantation, and the beautifully preserved house is occupied by the descendants of John D. Murrell.

Walnut and oak Benjamin chairs surround Tally-Ho's oval walnut dining table, which seats fourteen. The centerpiece, of porcelain by Boehm, depicts a pair of mourning doves and is flanked by Ormolu candelabra. On the Hunt table is an imported Chinese punch bowl. The antique highchair dates from the late 1800s. Over the fireplace hangs an early-1800s still life by George Lance of the English Royal Academy School.

The children's bedroom at Tally-Ho has twin half-tester iron beds and a folding iron crib from France.

Tezcuco

Near Darrow

Construction of Tezcuco Plantation Home was begun in 1855 and completed in 1860 by Benjamin F. Tureaud and his wife, Aglae Bringier, the original owners. This prosperous plantation remained in the possession of their descendants until 1946 when it was purchased by the Rouchon family.

It was built in the decade preceding the Civil War when antebellum Greek Revival architecture reached the height of interior embellishment. The ceiling cornices and center rosettes in this house have marvelously executed plaster detail, and all the interior doors and window sashes retain the original false graining, or *faux bois,* which was painstakingly painted by hand.

Many priceless antiques, impressive mahogany and rosewood pieces made by the famous New Orleans cabinetmakers, Mallard & Seignouret, enrich the interior.

Tezcuco is a French Creole home built of cypress and bricks from the plantation land. Six wooden columns extend across the recessed front gallery supporting the elaborately carved entablature. Columns of delicately carved iron filigree add to the beauty of the side gallery. The hipped roof has six dormer windows.

In 1950 Dr. and Mrs. Robert Potts bought this historic home, and it remained in their possession until 1982 when it was purchased by General and Mrs. O. J. Daigle. Through extensive restoration, the Daigles have returned the mansion to the grandeur characteristic of River Road plantation homes.

The striking French pier mirror dominates this view of the parlor at Tezcuco. The antique rosewood furniture was made by Seignouret. In the center of the marble-topped table an exquisite porcelain horse-drawn carriage attracts attention.

Four unusual antique lamps light the parlor at Tezcuco. The piano was made of rosewood in the 1800s. Note the Baccarat crystal chandelier.

The massive mahogany sideboard, the Tiffany chandelier, and an enormous mirror, original to Tezcuco, enhance the elegance of the long dining room.

The master bedroom at Tezcuco contains this four-poster Mallard bed with full tester and an elaborate oval mirror over the black cast-iron mantelpiece.

On the grounds of Tezcuco stands this small chapel. The gazebo to the side of the chapel houses the plantation bell.

One has a feeling of serenity upon entering the quiet chapel on the grounds of Tezcuco.